G000022449

# ALL FLUID-FLOW-REGIMES SIMULATION MODEL FOR INTERNAL FLOWS

# ENGINEERING TOOLS, TECHNIQUES AND TABLES

Additional books in this series can be found on Nova's website under the Series tab.

ENGINEERING TOOLS, TECHNIQUES AND TABLES

# ALL FLUID-FLOW-REGIMES SIMULATION MODEL FOR INTERNAL FLOWS

J.P. ABRAHAM
E.M. SPARROW
W.J. MINKOWYCZ
R. RAMAZANI-REND
AND
J.C.K. TONG

Nova Science Publishers, Inc.
*New York*

For permission to use material from this book please contact us:
Telephone 631-231-7269; Fax 631-231-8175
Web Site: http://www.novapublishers.com

## NOTICE TO THE READER

Additional color graphics may be available in the e-book version of this book.

### LIBRARY OF CONGRESS CATALOGING-IN-PUBLICATION DATA

All fluid-flow-regimes simulation model for internal flows / authors, J.P. Abraham ... [et al.].
    p. cm.
  Includes bibliographical references and index.
  ISBN 978-1-61122-503-7 (softcover)
  1. Pipe--Fluid dynamics. 2. Laminar flow--Mathematical models. I. Abraham, J. P. (John P.)
  TJ935.A46 2010
  620.1'064015118--dc22
                        2010041331

*Published by Nova Science Publishers, Inc. † New York*

# CONTENTS

# PREFACE

A multi-regime fluid flow model for internal flows has been applied to several pipe and duct problems. The investigated flow regimes and inter-regime transformations include fully laminar and fully turbulent, laminarization, and turbulentization. The model auto-selects both the flow regimes and the inter-regime transformations. It was employed here to study both steady and unsteady flows as well as flows in pipes and ducts of both axially unchanging cross section and axially enlarging cross section.

The local nature of the flow was quantified by two related metrics. One of these is the laminarization parameter which is the ratio of the rates of turbulence production and turbulence destruction. The other is the intermittency, whose role is to dampen the rate of turbulence production at locations where the flow is not fully turbulent. Among the key results, fully developed friction factors in transition regimes, heretofore unpredicted in the literature, have been determined and presented. Heat transfer was also investigated in certain of the aforementioned fluid-flow problems. A major issue in the prediction of heat transfer coefficients in regions of flow transition is the need for a connection between the turbulent viscosity and the turbulent thermal conductivity.

# NOMENCLATURE

| | |
|---|---|
| $a$ | transitional model constant |
| $A$ | transitional model constant |
| $c_p$ | specific heat |
| $d, \quad D$ | pipe diameters |
| $E$ | intermittency destruction term |
| $h$ | heat transfer coefficient |
| $H$ | channel height |
| $f$ | friction factor |
| $F_1, F_2$ | blending functions in SST model |
| $k$ | thermal conductivity |
| $Nu$ | Nusselt number |
| $p$ | pressure |
| $P$ | model production terms |
| $Pr$ | Prandtl number |
| $R$ | pipe radius |
| $Re$ | Reynolds number based on pipe diameter or hydraulic diameter |
| $S$ | absolute value of the shear strain rate |
| $t$ | time |
| $T$ | temperature |
| $u_i$ | local velocity |
| $U$ | average velocity |
| $x_i$ | tensor coordinate direction |

## GREEK SYMBOLS

| | |
|---|---|
| $\beta_1, \beta_2$ | SST model constants |
| $\varepsilon$ | turbulence dissipation |
| $\pounds$ | laminarization parameter |
| $\omega$ | specific rate of turbulence dissipation |
| $\mu$ | dynamic viscosity |
| $\kappa$ | turbulent kinetic energy |
| $\Pi$ | intermittency adjunct function |
| $\sigma$ | Prandtl-like diffusivities |
| $\gamma$ | intermittency |
| $\rho$ | density |
| $\theta$ | diffuser divergence angle |
| $\phi$ | turbulent Prandtl-number function |
| $\tau$ | harmonic period |

## SUBSCRIPTS

| | |
|---|---|
| $i,j$ | tensor indices |
| $fd$ | fully developed |
| $turb$ | turbulent |
| $UHF$ | uniform heat flux |
| $UWT$ | uniform wall temperature |
| $\gamma$ | intermittency |
| $\omega$ | specific rate of turbulence dissipation |
| $\Pi$ | intermittency adjunct function |
| $\kappa$ | turbulent kinetic energy |

*Chapter 1*

# INTRODUCTION

Although it is virtually the standard to analyze fluid flows for preselected unique flow regimes, it is very common in practice that a flow undergoes transitions between regimes. These transitions may begin in the laminar state, pass through intermittency, and become turbulent. Alternatively, a flow may be turbulent in a certain part of a ducting system but may transist into intermittency and eventually laminarize. To cope with such transitions, it is necessary to develop a solution method that will handle the various regimes and transitions automatically. It is clear that with sufficient computational resources, direct numerical simulation (DNS) is capable of dealing with such flows [1-9]. However, the resources required are uncommonly available, especially in industrial settings. There remains a practical need for a fluid-flow model which is capable of continuously and automatically identifying flow regimes and implementing their solution. In this review, such a method is described and applied to a variety of fluid flow situations.

The situations to be set forth here to illustrate the capability of the method include flows in pipes and ducts whose cross sections are axially unchanging, duct flows in which there is a change of cross section, and applications where the flow may be unsteady. These situations are studied here from both the standpoints of fluid-flow and heat-transfer phenomena, The latter encompass the two most standard thermal boundary conditions, uniform wall heat flux (UHF) and uniform wall temperature (UWT). In the presentation of results, focus will be directed to those features that are unique with respect to flow regime transitions.

Studies of laminar-turbulent transition date back to the classical experiment of Osborne Reynolds in 1883 [10]. It appears that the first quantitative study of intermittency, which was then defined as the fraction of time that the flow at a given point was turbulent, originated with Emmons in 1951 [11] and Mitchener in 1954 [12]. Emmons related the probability that the flow is turbulent at a given point to the intermittency. This concept was extended and elucidated by Dhawan and Narasimha in 1958 [13]. In a pair of papers [14, 15], Libby set forth an intermittency factor used at the interface between turbulent shear flows and adjacent irrotational flows. That intermittency factor was assigned a value of one in the turbulent region and a value of zero in the irrotational flow. In 1982, Patel and Scheuerer [16] defined an eddy viscosity which consisted of the $\kappa$–$\varepsilon$ eddy viscosity definition multiplied by the intermittency factor. This use of the intermittency was continued in work which appeared in the subsequent literature.

A seminal review of both theoretical and experimental work in this area, up to 1990, was provided by Mayle [17]. In a series of papers, Suzen and co-workers presented a one-equation model for the intermittency as a supplement of the Shear Stress Transport (SST) turbulence model [18-21]. That intermittency factor was used as a mulitiplier of the SST-defined eddy viscosity. Although this approach was capable of yielding useful results, its structure was ill-suited for modern, multi-processor computational schemes. The difficulty is that it requires integrated boundary layer parameters for its implementation rather than local parameters. The capture of these integrated parameters cannot be performed in a parallel-processor environment.

The most current and readily implemented scheme for predicting transition and skin friction in external flows is that formulated by Menter and co-workers [22-24]. They devised a scheme in which two supplementary equations, one for the intermittency and the other for the transitional Reynolds number, were employed in conjunction with the SST turbulence model. Two of the essential features of the Menter scheme are: (a) the intermittency is a multiplier of the rate of production of turbulence and (b) only local quantities appear in the formulation. It is relevant to note that the Menter intermittency factor differs from that of Suzen in that the latter used the intermittency as a multiplicative modifier of the eddy viscosity. Furthermore, the fact that only local quantities are used enables the Menter approach to be implemented on multi-processor computers. As with all eddy-viscosity-based models, adjustable constants appear both in the constitutive equations as well as multipliers in the turbulence production and destruction terms. In the Menter model, the constants were determined by comparisons with experimental data for external flows.

Computational experiments performed by the authors disclosed that Menter's model with its cadre of tuned constants was incapable of predicting the fluid flow in pipes and ducts [25]. In response to this finding, the constants were retuned so that the model provided results that agreed with accepted information for pipe flows. In particular, fully developed friction factors were used as the standard for the tuning activity. A difficulty arises when the transition model is applied to heat transfer. In conventional turbulent-flow-heat-transfer simulations, the process for the obtainment of the turbulent thermal conductivity is to use the already-obtained result for the turbulent viscosity and to employ the turbulent Prandtl number to extract the corresponding conductivity value. This procedure is, clearly, inappropriate when the flow is not fully turbulent. As a consequence, a provisional turbulent-Prandtl-number dependence on the Reynolds number was formulated by means of a tuning approach for which established Nusselt numbers in the transition regime were employed as the standard.

The transition model as described in the foregoing has been employed for a wide range of physical situations which include:

### Steady flows:

a)  Flow in a constant-area pipe with a single inlet velocity profile and turbulence intensity but with arbitrary inlet Reynolds numbers

b)  Flow in a parallel-plate channel with various inlet velocity conditions encompassing profile shape, turbulence intensity, and Reynolds number

c)  A piping system in which there is a conical enlargement of various expansion angles connecting upstream and downstream pipes of constant cross section with arbitrary Reynolds numbers

### Unsteady flows:

d)  Harmonic time-dependent flows

Schematic diagrams of these various physical situations are displayed in Figure 1.

The description of the simulation method and the relevant governing equations will now be presented.

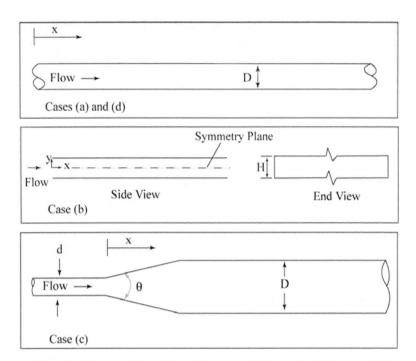

Reproduced with permission from *International Journal of Heat and Mass Transfer*, (in press).

Figure 1. Diagrams of the cases to be investigated.

# GOVERNING EQUATIONS

The equation statement begins with the customary RANS version of the Navier-Stokes equations. In the RANS model, the turbulence is assumed to be isotropic and fully embodied in a single quantity, the turbulent viscosity $\mu_{turb}$. To provide a means for the determination of $\mu_{turb}$, several two-equation models have been proposed in the 40-year span since the formulation of the $\kappa$–$\varepsilon$ model of Spalding and Launder [26-27]. In more recent times, the isotropy assumption has been lifted by the creation of new models, including the Reynolds-stress model (RSM) [28], large-eddy simulation (LES) [29], and direct numerical simulation (DNS) [1-9]. The first two of these do not have, to the best knowledge of the authors, a transition model. With regard to DNS, the issue of transition is moot. However, the computational assets needed for the accurate execution of DNS solutions are well beyond those typically available. In light of this survey, the authors have deemed the use of the enhanced Menter model as the most serviceable method for internal flow situations.

There are four sets of equations that must be dealt with to implement the model. The first set encompasses the conservation of mass equation and the three equations that constitute the RANS model. The next set serves to implement the RANS model by providing needed values of the turbulent viscosity. Among the numerous two-equation models that are available, the Shear Stress Transport model (SST) is necessarily the chosen one because it is mated with a transition model. The third set is the transition model itself. For heat transfer problems, the fourth set consists of energy conservation and a relationship between the turbulent Prandtl number and the Reynolds number. These equations are set forth in the following.

Mass conservation,

$$\frac{\partial u_i}{\partial x_i} = 0 \tag{1}$$

Reynolds-averaged Navier-Stokes equations (RANS),

$$\rho\left( u_i \frac{\partial u_j}{\partial x_i} \right) = -\frac{\partial p}{\partial x_i} + \frac{\partial}{\partial x_i}\left( (\mu + \mu_{turb}) \frac{\partial u_j}{\partial x_i} \right) \quad j = 1,2,3 \tag{2}$$

Shear Stress Transport model (SST),

$$\frac{\partial(\rho u_i \kappa)}{\partial x_i} = \gamma \cdot P_\kappa - \beta_1 \rho \kappa \omega + \frac{\partial}{\partial x_i}\left[ \left( \mu + \frac{\mu_{turb}}{\sigma_\kappa} \right) \frac{\partial \kappa}{\partial x_i} \right] \tag{3}$$

$$\frac{\partial(\rho u_i \omega)}{\partial x_i} = A\rho S^2 - \beta_2 \rho \omega^2 + \frac{\partial}{\partial x_i}\left[ \left( \mu + \frac{\mu_{turb}}{\sigma_\omega} \right) \frac{\partial \omega}{\partial x_i} \right] + 2(1-F_1)\rho \frac{1}{\sigma_{\omega 2}\omega} \frac{\partial \kappa}{\partial x_i} \frac{\partial \omega}{\partial x_i} \tag{4}$$

The solution of Eqs. (3) and (4) yields the values of $\kappa$ and $\omega$, which are then used to evaluate the turbulent viscosity $\mu_{turb}$ from

$$\mu_{turb} = \frac{a\rho\kappa}{\max(a\omega, SF_2)} \tag{5}$$

Transition model,

$$\frac{\partial(\rho\gamma)}{\partial t} + \frac{\partial(\rho u_i \gamma)}{\partial x_i} = P_{\gamma,1} - E_{\gamma,1} + P_{\gamma,2} - E_{\gamma,2} + \frac{\partial}{\partial x_i}\left[\left(\mu + \frac{\mu_{turb}}{\sigma_\gamma}\right)\frac{\partial\gamma}{\partial x_i}\right] \tag{6}$$

and

$$\frac{\partial(\rho\Pi)}{\partial t} + \frac{\partial(\rho u_i \Pi)}{\partial x_i} = P_{\Pi,t} + \frac{\partial}{\partial x_i}\left[\sigma_{\Pi,t}(\mu + \mu_{turb})\frac{\partial\Pi}{\partial x_i}\right] \tag{7}$$

Energy conservation,

$$\rho c_p\left(u_i \frac{\partial T}{\partial x_i}\right) = \frac{\partial}{\partial x_i}\left((k + k_{turb})\frac{\partial T}{\partial x_i}\right) \tag{8}$$

Turbulent Prandtl number,

$$\text{Pr}_{turb} = \frac{c_p \mu_{turb}}{k_{turb}} = \phi(\text{Re}) \tag{9}$$

The function of $\phi(\text{Re})$ is displayed in Figure 2.

The many symbols contained in these equations are defined in the nomenclature. Of particular note is the intermittency $\gamma$. It serves as the multiplier of the turbulence production term $P_\kappa$ in Eq. (3). Since $\gamma \leq 1$ in regions of flow transition, it damps the production of turbulence in those regions.

Equations (1) - (7) require simultaneous solution. In principle, the solution of Eqs. (8) and (9) can be performed after the determination of the velocities and the turbulent viscosity; however, in practice, it was found more convenient to solve all of Eqs. (1) – (9) simultaneously. The numerical work associated with the solution of the foregoing governing equations was performed by means of ANSYS CFX 12.0.

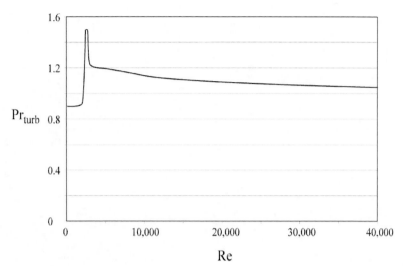

Reproduced with permission from *International Journal of Heat and Mass Transfer*, 52, 557-563, (2009).

Figure 2. Turbulent Prandtl number spanning the transition and fully turbulent-flow regimes deduced from experimental data for air.

*Chapter 2*

# FLOW IN A CONSTANT-AREA PIPE

The first application of the enhanced Menter model is fluid flow in a round pipe with arbitrary inlet Reynolds numbers. One of the goals of this investigation is to demonstrate the robustness of the model by exhibiting its predictions in all flow regimes while automatically identifying the nature of the regime. The numerical implementation of Eqs. (1)-(7) was accomplished by use of a carefully deployed mesh consisting of approximately 500,000 elements which were selected after the completion of a careful mesh-independence study. The most critical feature of the mesh is the deployment of nodes that are very close to the bounding wall of the pipe. The distance of the nearest node was tracked by the dimensionless distance $y^+$. In particular, the requirement that $y^+ < 1$ was strictly enforced.

The inlet Reynolds number was varied over the range from 1500 to 10,000. In all cases considered, the inlet velocity profile was assumed to be uniform with a turbulence intensity of 5%. The length of the pipe was x/D = 200, which was chosen to ensure that fully developed conditions were actually achieved.

## EVIDENCE OF LAMINARIZATION
## AND TURBULENTIZATION

The first result, to be presented in Figure 3, is the variation of the axial pressure gradient $dp/dx$ as a function of position along the length of the pipe. The choice of this quantity to initiate the results presentation is that it vividly illustrates the laminarization phenomenon that sets in immediately downstream of the inlet and the turbulentization that follows, leading to a fully developed turbulent flow.

Inspection of Figure 3 reveals a pattern which is distinctly different from that which is well established for flows that remain in a single regime. The steep drop in pressure gradient that occurs just downstream of the inlet reflects both high shear and momentum increase that normally occurs in the entrance region. However, what is unique is the attainment of a deep minimum followed by a sharp rise to an ultimately constant value of the pressure gradient. It is the sharp reversal in the magnitude of the pressure gradient that signals a flow regime transition. The ultimate flow regime is turbulent, from which it follows that the preceding regime is laminar. It further follows that the initially turbulent flow at the inlet experiences laminarization immediately upon entering the pipe. These findings are consistent with occurrences on a flat plate. In the absence of turbulators at the leading edge of the plate, the flow in the boundary layer in the upstream portion of the plate will always be laminar.

Also plotted in Figure 3 is a curve representing the variation of the pressure gradient predicted by the SST turbulence model without the transition supplement. The nature of the two pressure-gradient distributions is distinctly different. The slight local depression in the curve for the non-transition model is not an artifact; rather, it has been encountered experimentally [30]. The significant difference between the two curves displayed in Figure 3 is clear evidence of the workings of the transition model.

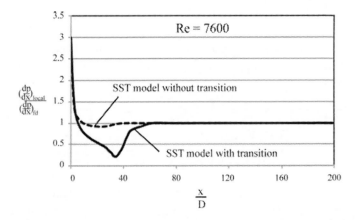

Figure 3. Axial distributions of the streamwise pressure gradient predicted by the SST model supplemented with transition compared with that from the SST model without transition.

To underscore the interpretation presented in the preceding paragraph, the behavior of the intermittency $\gamma$ will now be presented and discussed. The $\gamma$ results are shown in Figure 4 where radial profiles of $\gamma$ at various axial stations are displayed. The laminarization that occurs downstream of the pipe inlet can be observed in the figure. The first of the $\gamma$ profiles is at $x/D = 1$. That profile reflects the turbulence carried into the pipe. With increasing downstream distance, laminarization is clearly in evidence (decreasing $\gamma$) in the near-wall region and continues until approximately $x/D \sim 20$. Thereafter, turbulence recovers, and $\gamma$ begins to grow in the near-wall region, terminating with the fully developed $\gamma$ profile.

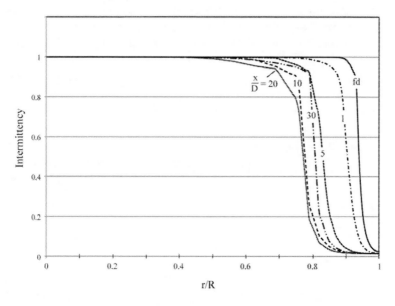

Figure 4. Radial distribution of the intermittency at various axial locations for an inlet turbulence intensity of 5%.

## FULLY DEVELOPED FRICTION FACTORS

With regard to practical applications, the most important quantity is the fully developed friction factor. Inspection of the classical literature, such as the Moody Chart, conveys results for pure laminar and pure turbulent flow but leaves a gap for the transition region. Here, by means of the transition model, that gap has been filled in a definitive manner as exhibited in Figure 5. That figure shows asymptotic lines for the aforementioned flow regions, and the new computational results are depicted as individual data points. The computed results are in excellent agreement with both the laminar and turbulent asymptotes. However, of greatest significance is the logical bridging between the asymptotes that has been provided by the current transition model. The equation of the bridging curve is

$$f_{fd} = 3.03 \times 10^{-12} \cdot \text{Re}^3 - 3.67 \times 10^{-8} \cdot \text{Re}^2 + 1.46 \times 10^{-4} \cdot \text{Re} - 0.151 \tag{10}$$

$$2300 < \text{Re} < 4500$$

This bridging equation and its relationship with the asymptotic lines suggest the existence of a transition flow regime that exists in the Reynolds number range between 2300 and 4500. In that range, the state of the flow is still intermittent but it is also fully developed. In this light, the new flow regime has been termed the *fully developed intermittent regime.* The availability of Eq. (10) now enables practical heat-exchanger design to be carried out on a more logical basis than in the past. The special relevance of this outcome is that a great many heat exchangers actually operate in the transition regime.

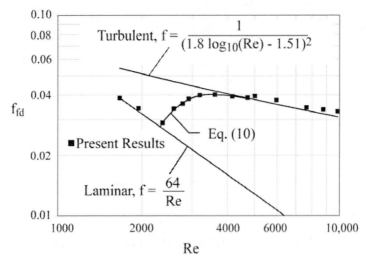

Reproduced with permission from *Numerical Heat Transfer B*, 54, 103-115, (2008).

Figure 5. Fully developed friction factors predicted by the new transition model and comparison with the literature asymptotes for laminar and turbulent flow.

The application of the new results obtained here is most effective when the regions of the pipe occupied by the different flow regimes are known.

This information is conveyed in Figure 6. The figure shows two lines which respectively delineate the locations of laminar flow breakdown and of the onset of the fully developed regime. The latter curve does not clearly indicate the type of the fully developed regime that is attained. For the Reynolds number range between 2300 and 4500, the fully developed regime is intermittent while for Reynolds numbers above 4500, the fully developed regime is turbulent.

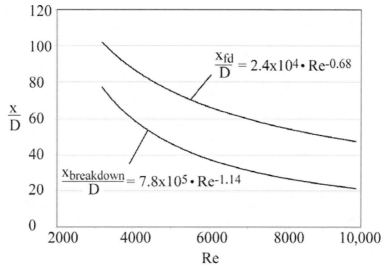

Reproduced with permission from *Numerical Heat Transfer B*, 54, 103-115, (2008).

Figure 6. Locations of laminar flow breakdown and of the establishment of fully developed flow as functions of Reynolds number for an inlet turbulence intensity of 5% and a flat velocity profile.

*Chapter 3*

# FLOW IN A PARALLEL-PLATE CHANNEL

The study of transitioning flows in parallel-plate channels was motivated not only to verify that the new transition model is effective for non-circular internal flows, but also to investigate the impact on transition of various shapes of the inlet velocity profile and various levels of initial turbulence intensity [31]. In particular, both flat and parabolic inlet profiles were investigated while the turbulence intensity ranged from 1-5%. The governing equations for this problem are the same as those already stated in Eqs. (1) – (7). Aside from the variants of the inlet conditions, the boundary conditions for the present problem are the same as those for the foregoing round-pipe study. The number of elements used for the discretization was approximately 350,000 and mesh independence was attained.

The cases investigated here are listed in Table 1. It is seen that for each inlet profile shape, two turbulence intensities, 1 and 5%, were imposed.

## Table 1. Identification of the investigated cases

| Case | Inlet Velocity Profile | Inlet Turbulence Intensity (%) |
|------|------------------------|-------------------------------|
| A | Parabolic | 5 |
| B | Flat | 5 |
| C | Parabolic | 1 |
| D | Flat | 1 |

# EVIDENCE OF LAMINARIZATION
# AND TURBULENTIZATION

To provide evidence of the various flow transitions that may occur, it was deemed appropriate to plot the streamwise variation of the centerline velocity in the dimensionless form $u_{max}/U$, where $U$ is the axially unchanging cross-sectional mean velocity. Figure 7 is the first of a sequence of figures, each corresponding to a case listed in Table 1 and specifically to Case A. That case is characterized by a parabolic velocity profile at inlet and a 5% inlet turbulence intensity. The figure displays a kaleidoscope of trends. For all the cases exhibited in the figure, $u_{max}/U = 1.5$ at the inlet. The Reynolds number is based on the cross-sectional average velocity and the hydraulic diameter.

Inspection of the figure shows that the curves separate themselves into two groups according to their trends with the axial position $x/D$. Those curves that correspond to Reynolds numbers below 10,000 demonstrate an initial steep drop, the attainment of a minimum, and a recovery which terminates in an axially unchanging value of the centerline velocity. This behavior is, in part, due to the incompatible specification of the inlet conditions: a laminar velocity profile and a turbulence intensity that is different from zero. This combination results in the rapid drop of $u_{max}/U$ from its initial value of 1.5. That drop occurs as the turbulence intensity asserts itself. In that regard, it is worthy of note that the ratio $u_{max}/U$ for turbulent flows rarely exceeds 1.2. The minimum can be attributed to the fact that the fully turbulent regime is not attained in the fully developed portion of the flow as evidenced by the fact that the values of $u_{max}/U$ in that region are well above 1.2. This fact indicates that the fully developed regime for Reynolds numbers less than 10,000 is characterized as laminar or intermittent. At still higher Reynolds numbers, which corresponds to the second group of curves, the fully developed regime is turbulent as witnessed by the fact that the $u_{max}/U$ values are in the range of 1.12-1.17. As a consequence, there is no recovery of the $u_{max}/U$ curves for those Reynolds numbers.

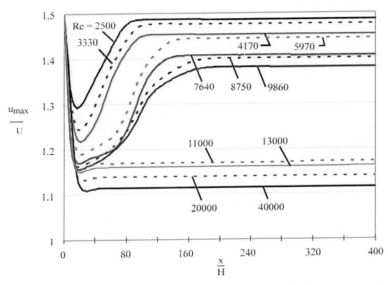

Reproduced with permission from *International Journal of Heat and Mass Transfer*, 52, 4040-4046, (2009).

Figure 7. Streamwise variation of $u_{max}/U$ for a parabolic inlet velocity profile and an inlet turbulence intensity of 5% (Case A).

The next situation to be considered is Case B of Table 1, for which results are presented in Figure 8. That case is characterized by the same turbulence intensity (5%) as was Case A, but now, the velocity profile at the inlet is flat, as can be seen from the initial value $u_{max}/U = 1$. The instantaneous rise in the centerline velocity immediately downstream of the inlet can be attributed to the impossibility of maintaining a flat velocity profile in the presence of wall friction. Only at the lowest Reynolds number does the initial rise monotonically persist to the attainment of a fully developed flow. For Reynolds numbers in the range of 4000-5000, the curves display a local maximum as the endpoint of their initial rise, and this is followed by a modest decrease to a minimum and then a subsequent increase to a fully developed value. The first maximum can be attributed to the impetus toward fully turbulent flow provided by the initial turbulence intensity of 5%. This impetus is not sufficient for the attainment of fully developed flow since the Reynolds numbers are not high enough. This insufficiency gives rise to the moderate decrease of $u_{max}$.

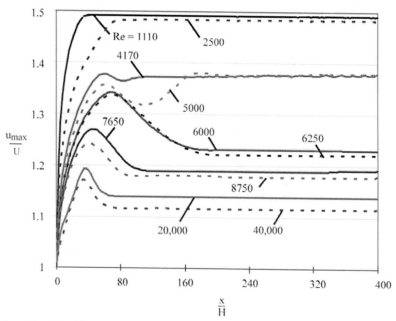

Reproduced with permission from *International Journal of Heat and Mass Transfer*, 52, 4040-4046, (2009).

Figure 8. Streamwise variation of $u_{max}/U$ for a flat inlet velocity profile and an inlet turbulence intensity of 5% (Case B).

The next case to be considered is Case C of Table 1, the results for which are presented in Figure 9. The conditions describing this case are somewhat inconsistent in that they consist of a fully developed laminar velocity profile and a 1% turbulence intensity at the inlet. The results shown in the figure naturally subdivide themselves into three groups. The uppermost group corresponds to Reynolds numbers which would normally give rise to fully developed laminar flows. There is a moderate development length within which the initial turbulence level is dissipated. The intermediate group is somewhat unique in that the fully developed regime is characterized by an intermittent flow. That region has been designated as *fully developed intermittent*. Of particular interest in this group is the behavior of the case Re = 9860 which displays an initial decrease which is reminiscent of the pattern expected for a flow whose ultimate fate is fully turbulent. However, full turbulence is not achieved and the flow reverts to intermittency. The lowermost set of curves pertains to Reynolds numbers that yield fully developed turbulent flow. Those cases correspond to fully developed $u_{max}/U$ values in the range of ~1.12-1.16.

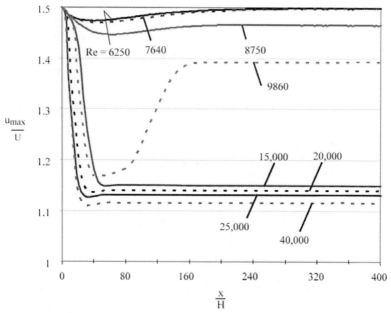

Reproduced with permission from *International Journal of Heat and Mass Transfer*, 52, 4040-4046, (2009).

Figure 9. Streamwise variation of $u_{max}/U$ for a parabolic inlet velocity profile and an inlet turbulence intensity of 1% (Case C).

The last situation to be discussed is Case D of Table 1, and the corresponding results are displayed in Figure 10. This case is for a flat velocity profile and a turbulence intensity of 1% at the inlet. For those Reynolds numbers which would normally lead to a fully developed laminar flow, the initial turbulence intensity is readily dissipated and laminar conditions are retained without perturbation. The terminal velocity ratio $u_{max}/U = 1.5$ is that of a fully developed laminar flow. The lower group of curves encompasses Reynolds numbers which, when given ample development length, give rise to fully turbulent flows. However, there is a long development length within which the initial laminarization is dissipated and turbulence eventually triumphs.

Attention will now be turned to the fully developed friction factor. Of particular interest is the identification of how this quantity responds to the inlet conditions expressed by the shape of the velocity profile and the magnitude of the turbulence intensity. This information is conveyed in Figure 11 where the results for all of the four cases listed in Table 1 are presented.

Among the reference lines, that labeled 96 Re$^{-1}$ corresponds to laminar flow in a parallel-plate channel. A second reference line labeled 0.507 Re$^{-0.3}$ is an experimental correlation for low-Reynolds-number turbulent flows [32]. This reference line mates perfectly with the well-known Colebrook equation, labeled as [1.8log(Re/6.9)]$^{-2}$, which is believed to be the best fit for turbulent friction factor data over a very wide range of Reynolds numbers [33].

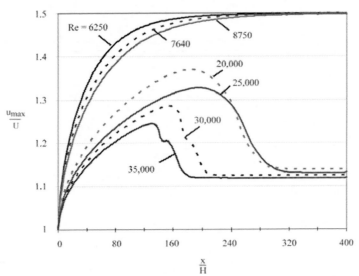

Reproduced with permission from *International Journal of Heat and Mass Transfer*, 52, 4040-4046, (2009).

Figure 10. Streamwise variation of $u_{max}/U$ for a flat inlet velocity profile and an inlet turbulence intensity of 1% (Case D).

An overview of Figure 11 indicates that Case B displays a behavior which is similar to that displayed on traditional plots of friction factor versus Reynolds number. Specifically, the friction factor results for that case coincide with those for laminar flow up to a Reynolds number of approximately 2300 and subsequently bridge smoothly across a transition region which terminates in a merging with the turbulent flow correlations at a Reynolds number of approximately 8000. This breakdown Reynolds number is in good accord with experimental data reviewed in the Introduction. The resulting new correlation for Case B, in the range between $2300 < Re < 8000$, is

$$f_{fd} = -3.82 \times 10^{-13} \cdot Re^3 + 7.40 \times 10^{-9} \cdot Re^2 - 4.40 \times 10^{-5} \cdot Re + 0.109 \quad (11)$$

On the other hand, the results for Cases A, C, and D display a rather different behavior. For those cases, the friction factor results are nearly coincident with those for laminar flow up to Reynolds numbers of 8000 - 10,000, at which point they break sharply upward and mate with the turbulent results, apparently without an extended transition regime.

It is believed that the mode of laminar breakdown for Cases A, C, and D is different from that of Case B. For the former, the breakdown is due to the instability of a fully developed profile to fluctuations inherent in the prescribed turbulence intensity. Cases A and C have imposed parabolic velocity profiles at the inlet; Case D achieves a nearly parabolic profile as the flow develops in the streamwise direction. In contrast, the behavior of the friction factors for Case B is similar to that which occurs in a boundary layer flow.

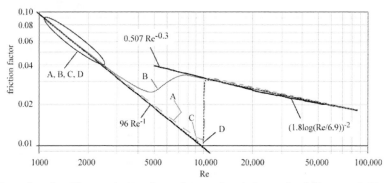

Reproduced with permission from *International Journal of Heat and Mass Transfer*, 52, 4040-4046, (2009).

Figure 11. Fully developed friction factors for flow in a parallel plate channel.

# Chapter 4

# FLOW IN A DIFFUSER WITH UPSTREAM AND DOWNSTREAM STRAIGHT PIPE SECTIONS

A diffuser is a section of piping in which the diameter increases in the streamwise direction. The presence of the diffuser causes a reduction of the Reynolds number of the flow which may trigger a subsequent transition.

Depending on the Reynolds number of the flow that enters the diffuser, the flow may undergo a change of regime. In particular, if the entering Reynolds number is either in the turbulent or intermittent regime, the exiting flow might either be laminar or intermittent. This kaleidoscope of possible outcomes will be investigated in detail. The physical situation to be considered is pictured in Figure 1(c), and the detailed parametric values of the Reynolds numbers and the diffuser angles are listed in Table 2. For all of the investigated cases, a fully developed velocity profile existed at the inlet of the diffuser [34-35].

## INTRODUCTION OF THE LAMINARIZATION PARAMETER

In a previous section of the paper, evidence of laminarization was offered by a display of the intermittency and its spatial variation. Here, a more general metric called the *laminarization parameter* £ is introduced that will serve as a measure of both laminarization and turbulentization. The definition of this metric is

$$£ = \frac{\text{rate of turbulence production}}{\text{rate of turbulence dissipation}} = \frac{\gamma \cdot P_k}{\varepsilon} \tag{12}$$

In this definition, the quantity $P_k$ is the rate of production of turbulence kinetic energy, and $\gamma$ plays the role of a damping factor in the case of intermittent or laminar flow. It may be expected that the laminarization factor would decrease during the process of laminarization and increase as turbulentization occurs.

**Table 2. Identification of the investigated cases**

| Case No. | Diffuser Inlet Re | Diffuser Outlet Re | Diameter Ratio | Divergence Angle |
|---|---|---|---|---|
| 1 | 500 | 125 | 4 | 5 |
| 2 | 1000 | 250 | 4 | 5 |
| 3 | 2000 | 500 | 4 | 5,10 |
| 4 | 8000 | 2000 | 4 | 5,10,30 |
| 5 | 6000 | 1500 | 4 | 5,10,30 |
| 6 | 4000 | 1000 | 4 | 5,10,30 |
| 7 | 14240 | 3560 | 4 | 5,10,30 |
| 8 | 12460 | 3115 | 4 | 5,10,30 |
| 9 | 11080 | 2770 | 4 | 5,10,30 |
| 10 | 20000 | 5000 | 4 | 5,10,30 |
| 11 | 33200 | 8300 | 4 | 5,10,30 |
| 12 | 50000 | 12500 | 4 | 5,10,30 |
| 13 | 8000 | 2670 | 3 | 5,10,30 |
| 14 | 6000 | 2000 | 3 | 5,10,30 |
| 15 | 4000 | 1330 | 3 | 5,10,30 |
| 16 | 8000 | 4000 | 2 | 5,10,30 |
| 17 | 6000 | 3000 | 2 | 5,10,30 |
| 18 | 4000 | 2000 | 2 | 5,10,30 |

# FLOWS WHICH LAMINARIZE

The first case to be considered in the presentation of results is characterized by a diffuser inlet Reynolds number of 8000 and an exit Reynolds number of 2000. The streamwise variation of the laminarization parameter is conveyed in Figure 12. The results to be presented there are representative of all other cases in Table 2 where laminarization occurs. In the figure, the laminarization parameter at a number of streamwise locations is plotted as a function of the radial coordinate. In interpreting the figure, it should be noted that the exit of the diverging section is at $x/d = 54$ (where $d$

is the diameter at the inlet of the diffuser). All of the diverging curves in the main body of the figure pertain to axial locations within the diffuser proper. The curves for successive $x/d$ stations indicate a monotonic decreasing trend of the laminarization parameter with increasing penetration into the diffuser; clearly, the flow is laminarizing.

The inset at the upper right of the figure shows results in the straight pipe that follows the exit of the diffuser. Of a special note are the very low values of the laminarization parameter. However, the decrease of $£$ to its far downstream value is not monotonic, as evidenced by the successive curves for $x/d$ = 300, 500, and 700. This behavior is believed due to a local imbalance in the rates of decrease of the production and destruction of turbulence and does not correspond to a rejuvenation of turbulence, as evidenced by the low values of $£$ and by the friction factor results to be presented shortly.

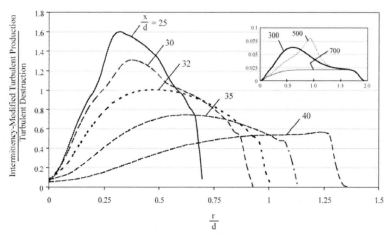

Reproduced with permission from *International Journal of Thermal Sciences*, 49, 256-263, (2010).

Figure 12. Radial and axial variations of the laminarization parameter for a diffuser angle of 5°. Diffuser inlet and exit Reynolds numbers of 8000 and 2000, respectively. Diffuser ends at $x/d$ = 54.

In the next figure, Figure 13, the results are displayed in a format similar to that of Figure 12. There are, however, major differences in that now the Reynolds number at the exit of the diffuser falls into the fully developed intermittent regime rather than in the laminar regime as was the case in Figure 12. Within the diffuser proper, laminarization is clearly taking place as the inlet Reynolds number of 14400 decreases monotonically. In the pipe downstream of the exit, the laminarization parameter increases with increasing downstream distances, thereby reversing the laminarization tendency that occurred within the diffuser proper. At far downstream distances, $x/d > 200$, an equilibrium distribution of the laminarization parameter is established and does not change with increasing downstream distances.

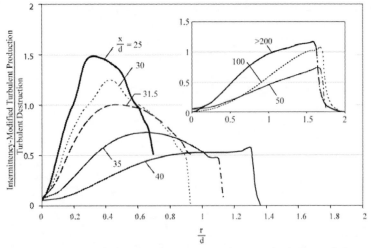

Reproduced with permission from *International Journal of Thermal Sciences*, 49, 256-263, (2010).

Figure 13. Radial and axial variations of the laminarization parameter for a diffuser angle of 5°. Diffuser inlet and exit Reynolds numbers of 14400 and 3600, respectively. Diffuser ends at $x/d = 54$.

## FULLY DEVELOPED FRICTION FACTORS

From the standpoint of practical utility, the friction factor is of greatest importance. Fully developed friction factor results are presented in Figure 14 as a function of the Reynolds number. The present results are displayed by the discrete data symbols. The three reference lines in the figure are (a) laminar fully developed flow, (b) turbulent fully developed flow, and (c) Eq. (10). The latter corresponds to a correlation presented earlier in this paper for fully developed flow in a pipe of unchanging cross section in which upstream changes of flow regime occurred. The congruence of the present results with those for pure laminar flow and pure turbulent flow is excellent.

Of particular interest is the fact that the present friction factors which span the transition regime are slightly different from those of Eq. (10). The difference between the present friction factors and those of Eq. (10) reflect the upstream history of the respective flows. In the present case, the flow experienced a laminarization caused by area change, and flow separation also occurred for many of the cases for which data are presented in the figure. In contrast, the upstream history of the flow that corresponds to Eq. (10) is a laminarization in a pipe of constant cross section, and no separation occurs in that case.

An algebraic equation was fitted to the present friction factor results in the range between Re = 2600 and Re = 5000. That equation is

$$f_{fd} = -3.73x10^{-9} \, Re^2 + 3.29x10^{-5} \, Re - 0.0319,$$

$$2600 < Re < 5000 \tag{13}$$

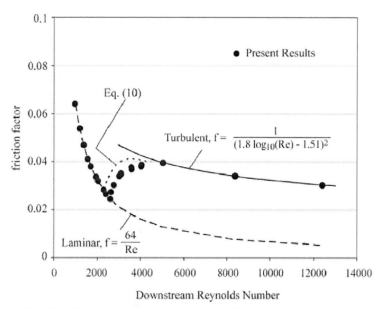

Reproduced with permission from *International Journal of Thermal Sciences*, 49, 256-263, (2010).

Figure 14. Variation of the fully developed friction factor with the Reynolds number.

# Chapter 5

# HARMONIC TIME-DEPENDENT FLOWS

The current great interest in time-varying flows can be traced to a number of critical applications. The range of motivating applications extends from Stirling power-producing cycles [36] to biological fluid flows [1, 4, 37-41] encompassing both respiration and the cardiac cycle. This type of flow is especially challenging in that the maximum and minimums that are encountered may well give rise to changes in flow regime. To model a typical flow of this type, a model problem has been developed and will be set forth here. That model is an incompressible flow in a round pipe with a harmonically varying inlet velocity. The timewise variation to be considered is displayed in Figure 15. The figure shows the time dependence of the Reynolds number in which the amplitude varies between 1000 and 5000. In this range, both laminarization and turbulentization will occur depending on the instantaneous value of the Reynolds number and on whether the flow is accelerating or decelerating.

Great care was exercised in the numerical implementation. The time step was selected to be $1/1000^{\text{th}}$ of the period of the imposed temporal variation. At each time step, 50 iterations of the governing equations were performed (inner loop iterations). A mesh-independence study was carried out with regard to the spatial discretization of the solution domain. Solutions were obtained using mesh sizes that ranged from 177,000 elements to approximately 600,000 elements. Not only was the number of elements varied, but also the deployment of the elements was carefully tailored. Furthermore, the node nearest the pipe wall obeyed the requirement that $y^{+} \sim$ 0.25. The outcome of the spatial mesh study demonstrated that friction results extracted from the numerical solutions agreed to within 0.5% for all of the meshes employed. A more detailed description of the numerical model is presented in [42].

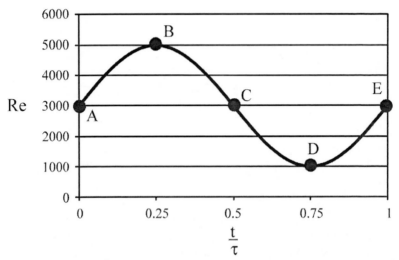

Reproduced with permission from *Numerical Heat Transfer Part A*, 56, 861-879, (2009).

Figure 15. Timewise variation of the cross-sectional average Reynolds number.

## FULLY DEVELOPED FRICTION FACTORS

To demonstrate the validity of the numerical simulation as well as to provide results of practical relevance, the friction factors extracted from the solution are presented in Figure 16. This figure contains a multitude of relevant information. In particular, it spans all possible flow regimes which are encountered during each cycle of the harmonic variation. In the early part of the cycle, the flow is in the process of becoming fully turbulent; that is, it is turbulentizing. This portion of the period corresponds to the range A-B in Figure 15. The second portion of the cycle, as depicted in Figure 16, is a fully turbulent regime which is centered about point B of Figure 15. Then, the next regime depicted in the figure represents flow becoming laminar (laminarizing). This regime is centered about point C in Figure 15. Finally, the last segment of the cycle is a fully laminar flow which is centered about point D in Figure 15.

Of particular note are the two curves in Figure 16. The solid lines are taken from steady state results which are applied in a quasi-steady manner.

On the other hand, the dashed lines represent the results of the present analysis. The agreement between the two sets of results is remarkable and lends very strong support both to the validity of the model and its implementation. The results presented in Figure 16 correspond to a period $\tau$ = 640 seconds. This lengthy period was chosen because it would be expected that the flow would behave as if it were quasi-steady. To support this expectation, Figure 17 has been prepared. In that figure, the fully developed friction factor is plotted as a function of time within a single period. Each of the curves that appear in Figure 17 corresponds to a different assigned period. From an appraisal of the results shown in the figure, it is clear that the results for $\tau$ = 640 seconds are terminal in the sense that the same results would be obtained for any $\tau$ greater than 640 seconds. This behavior demonstrates that the results at $\tau$ = 640 seconds are actually quasi-steady.

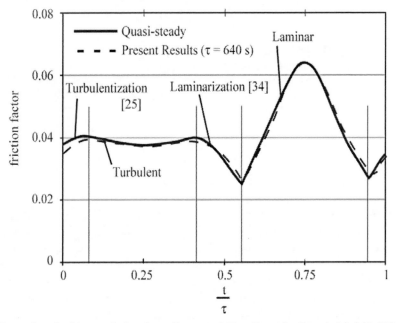

Reproduced with permission from *Numerical Heat Transfer Part A*, 56, 861-879, (2009).

Figure 16. Comparison of unsteady friction factor results with those obtained from a quasi-steady model.

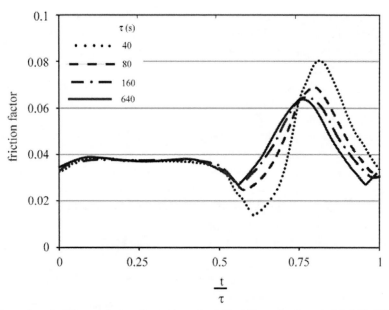

Reproduced with permission from *Numerical Heat Transfer Part A*, 56, 861-879, (2009).

Figure 17. Timewise variation of the fully developed friction factor with increasing oscillation period.

## EVIDENCE OF LAMINARIZATION
## AND TURBULENTIZATION

To provide encompassing evidences of laminarization and turbulentization in the time-dependent flow being considered, a format has been adopted in which the laminarization parameter and the corresponding velocity profiles are presented in a side-by-side display. The first of the figures to be shown is Figure 18. In the left-hand portion of the figure, the streamwise development of the laminarization parameter is exhibited. The graph for each axial station portrays a radial profile. In the right-hand portion of the figure, velocity profiles at a succession of axial stations are plotted. The particular instants of time for which these results apply are keyed to Figure 15 and correspond to a Reynolds number of 3000 (Instants A, C, and E).

It can be seen from Figure 18 that the laminarization parameter is moderately low in the immediate neighborhood of the pipe inlet. This outcome, which is consistent with what has already been demonstrated in Figure 3 is due to the viscous suppression of turbulence in the boundary layer just downstream of the pipe inlet. This suppression continues at moderate increases of the downstream distances, but, more importantly, the location of the maximum of the laminarization parameter shifts to smaller radial locations as the boundary layer thickens. At still larger downstream distances, for example, at $x/D = 50$, the near-inlet suppression of turbulence weakens and the laminarization parameter increases. As the flow approaches the fully developed regime, intermittency sets in, as witnessed by the attainment of laminarization parameters that exceed one.

Time instants A and E (Figure 15) are indicative of flows that are accelerating and are approaching the turbulent state. On the other hand, instant C corresponds to a flow that is decelerating and is in the process of laminarization. It can be seen from Figure 18 that the laminarization parameter values for these two situations are not identical. This finding suggests that the transition process has a memory of the previous state of the flow and that acceleration or deceleration affects the flow dynamics.

Figure 18 also shows a sequence of velocity profiles in which the accelerating flows (instants A and E) and the decelerating flow (instant C) are indistinguishable at axial locations up to $x/D \sim 50$. At larger downstream distances, the two cases tend to deviate slightly with the accelerating flow having a higher velocity in the neighborhood of the pipe axis. Of particular note is the centerline velocity and its numerical values. For a purely laminar fully developed flow, the ratio of the centerline velocity to the mean velocity is two, whereas for a turbulent flow, the ratio varies with the Reynolds number and has a value of approximately 1.26 for Re = 4000 and 5000. The topmost graph of Figure 18 exhibits centerline velocity ratios of 1.53 and 1.45, respectively, for the accelerating and decelerating situations. This finding is supportive of the present model in that the predicted centerline velocities fall between those for fully laminar and fully turbulent flows.

Information that is conveyed in Figure 19 corresponds to instant B in Figure 15 (Re = 5000). This figure clearly expresses the turbulent nature of the flow. At locations near the inlet ($x/D = 10$), the turbulence is moderated by the dampening effect of the laminar boundary layer. At $x/D = 25$, a double maximum distribution is seen. That distribution reflects the existence of turbulence in the boundary layer as well as at the interface of the boundary layer and the free stream. The remaining distributions are characterized by a single maximum whose peak value does not vary monotonically with $x/D$. This behavior is believed due to a local imbalance in the rates of decrease of the production and destruction of turbulence.

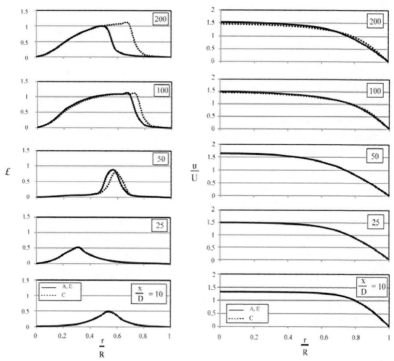

Reproduced with permission from *Numerical Heat Transfer Part A*, 56, 861-879, (2009).

Figure 18. Radial distributions of the laminarization parameter and the normalized streamwise velocity at various axial locations. Time instants A, C, and E are defined in Figure 15.

The development of the velocity profile appears to follow a normal pattern. However, closer inspection reveals that, among other features, the centerline velocity ratio does not increase monotonically with $x/D$ as would be expected by conventional wisdom. The non-monotonic behavior exhibited here is due to the accounting of processes that are often neglected in the literature. In particular, the fact that the entering flow experiences a laminar boundary layer is a key factor in both the non-monotonic behavior of the velocity ratio and the elongation of the velocity development length. Another relevant factor in the deviation of the present findings from those of conventional wisdom is that the flow being analyzed here is an unsteady flow whereas conventional wisdom is limited to steady flows. The development length that can be extracted from Figure 19 is approximately 50-100 diameters.

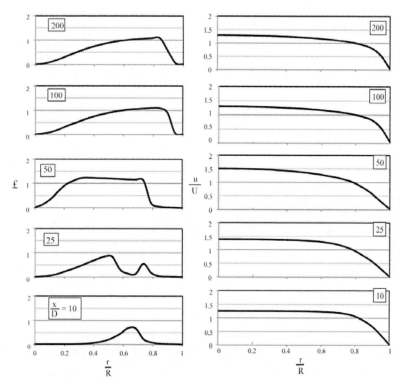

Reproduced with permission from *Numerical Heat Transfer Part A*, 56, 861-879, (2009).

Figure 19. Radial distributions of the laminarization parameter and the normalized streamwise velocity at various axial locations. The time instant B is defined in Figure 15.

The main message of Figure 20 (instant D, Re = 1000) is that laminar flows are characterized by very low values of the laminarization parameter at all axial locations. More careful evaluation of the figure indicates that, except for the immediate neighborhood of the inlet, the values of the laminarization parameter are less than 0.03. The relatively high value of the parameter at $x/D = 10$ is the result of the specified turbulence intensity of 5% at the inlet of the pipe.

The velocity profiles that are displayed in the right-hand column of the figure demonstrate the fully laminar nature of the flow. The near-inlet velocity profiles are relatively flat, reflecting the imposed flat-inlet-velocity condition. The growing boundary layer provides curvature to the profile with increasing downstream distance. At the greatest downstream distance, $x/D =$

200, the centerline velocity ratio is two in accordance with laminar-pipe-flow theory. If the onset of fully developed flow is defined as a 5% approach of the centerline velocity ratio to its limiting value of two, it then follows that development is achieved by $x/D = 50$, which is in accord with the simple and accepted laminar-flow entrance length $x/D = 0.05\ Re$.

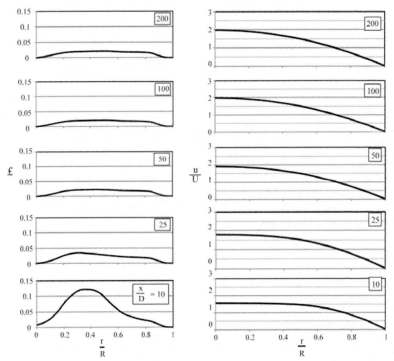

Reproduced with permission from *Numerical Heat Transfer Part A*, 56, 861-879, (2009).

Figure 20. Radial distributions of the laminarization parameter and the normalized streamwise velocity at various axial locations. The time instant D is defined in Figure 15.

*Chapter 6*

# HEAT TRANSFER RESULTS

Heat exchangers often operate under conditions where the flow is in the transition regime. That state of affairs has made the design of heat exchangers difficult to perform with a high degree of certainty. The results obtained here for the fluid flow can be used as a basis for the attainment of more accurate heat transfer coefficients in regimes of transition, and this task will now be addressed.

## THE TURBULENT PRANDTL NUMBER

There is a special issue which must be taken into account when heat transfer in regions of intermittent flow is being sought. For fully turbulent flows, a turbulent conductivity $k_{turb}$ is obtained by extending the available models which provide results for the turbulent viscosity $\mu_{turb}$. That extension is highly simplistic in that it is based on a proportionality provided by the turbulent Prandtl number, Eq. (9). A rearrangement of Eq. (9) yields

$$k_{turb} = \left(\frac{c_p}{Pr_{turb}}\right)\mu_{turb} \qquad (14)$$

It has been traditional to take the value of $Pr_{turb}$ as 0.9. Since $c_p$ is also a constant, the proportionality between $k_{turb}$ and $\mu_{turb}$ is thereby established. This simplistic approach cannot be carried over to regimes of transition. In preparation for the development of a model for determining $k_{turb}$ in regions of transition, a first step was a thorough search of the relevant literature.

A definitive source of information about the turbulent Prandtl number is a review paper by Kays [43]. Among other issues, the paper indicates that values of the turbulent Prandtl number in the near-neighborhood of a wall can substantially exceed the traditional value of 0.9. Crimaldi et al. [44] also measured high values of the turbulent Prandtl number; in particular, values as high as 10 for water flow. Other reports of relatively high values of $Pr_{turb}$ are given by Bensayah et al. [45] and by Chua et al. [46] for heat transfer in jets. In atmospheric boundary layers, it was found by Grachev et al. [47] and by Bass et al. [48] that turbulent Prandtl numbers strongly depend on variations of the Richardson number and can reach values in excess of 40. Heat transfer measurements in a ribbed channel provided values of the turbulent Prandtl number of approximately 1.7 in a gas flow [49]. In this light, the information that just cited suggests the need to rethink the standard default treatment of the turbulent Prandtl number which is typically a value of 0.9.

The choice of the values of the turbulent Prandtl number was made by comparing predictions of the fully developed Nusselt number with literature standards. Those predictions were made by solving Eqs. (1)-(9) for parametrically varied values of $Pr_{turb}$ for each of a succession of specified Reynolds numbers. The thermal boundary condition for this work was uniform wall heat flux because that was the condition employed in the experiments which served as the standard. The velocity boundary conditions for this work were a flat inlet profile with $Tu = 5\%$. Since the focus of the work was the attainment of fully developed results, the choice of the inlet conditions for the velocity is not expected to be of significance.

To implement the comparison between the predicted and experimental results, it is relevant to begin with the definitions of the fully developed heat transfer coefficient $h_{fd}$ and Nusselt number $Nu_{fd}$. They are

$$Nu_{fd} = \frac{h_{fd} D}{k}, \ h_{fd} = \frac{q}{\left(T_{wall} - T_{bulk}\right)}\bigg|_{fd} \tag{15}$$

The experimental data that were used to implement the aforementioned determination of the turbulent Prandtl number are displayed in Figure 21. That figure consists of two parts. The main part of the figure displays a large amount of fully developed experimental data for air flows in uniformly heated pipes, while the inset shows two popular algebraic correlations in the transition regime. For the range of Reynolds numbers between 4800 and 40,000, the experimental data served as the standard of comparison to enable the turbulent Prandtl number to be chosen in order that the predicted $Nu_{fd}$ values are congruent with the data. The faired curve that appears in the main part of the figure is to provide continuity for the data.

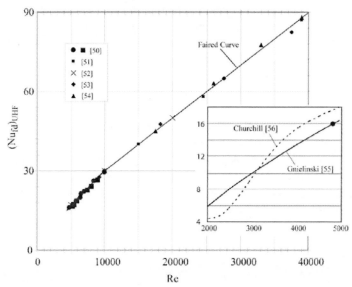

Figure 21. Available fully developed experimental Nusselt-number data and algebraic curve fits (inset) used for the determination of the turbulent Prandtl number.

Consideration of the most popular among the available algebraic correlations in the transition regime suggests some degree of insufficiency for both. The Gnielinski correlation [55] does not merge with the laminar results in a manner that would reinforce its acceptance for the lower-Reynolds-number portion of the transition regime. On the other hand, while the Churchill [56] correlation smoothly merges with the laminar regime, it clearly overestimates the experimental data at Re = 4800. Furthermore, its undulating behavior is inconsistent with rational expectations.

For the range of Reynolds numbers between 2300 and 4800, the turbulent Prandtl number determination was based on the Gnielinski [55] curve between the crossing-point Reynolds number of 3100 and 4800 and on the Churchill curve [56] for Reynolds numbers below 3100.

The values of the turbulent Prandtl numbers that resulted from the foregoing procedure are displayed in Figure 2 where they are plotted as a function of the Reynolds number. There is a sharp peak whose onset is at Re ~ 2300 and which attains its maximum at about 2500. Thereafter, the values of the turbulent Prandtl number decrease, sharply at first, and then more gradually before achieving a constant value of 1.05. At the peak, $Pr_{turb}$ equals 1.5. The Reynolds number range where the highest values of $Pr_{turb}$ occur is unique in that no literature information was unearthed. The intermittency which exists in that range is drastically different from that of a fully turbulent flow. From the foregoing literature survey, it was demonstrated that for situations that differ from those encountered in a fully developed pipe flow, values of $Pr_{turb}$ in excess of one are not uncommon. It is believed that until further studies of the nature of turbulence which occurs just after laminar breakdown have been carried out, the values displayed in Figure 2 should be considered as reasonable.

It is worthy of note that the $Pr_{turb}$ values exhibited in Figure 2 were found to be applicable for the two standard thermal boundary conditions, UHF and UWT, both of which were employed during the course of this study. This was earlier noted by Churchill [57].

## FULLY DEVELOPED NUSSELT RESULTS

The application of Eqs. (1)-(9) supplemented by the information conveyed in Figure 2 enabled the determination of fully developed Nusselt numbers for both the UWT and UHF boundary conditions over the entire range of Reynolds numbers. These results will be reported shortly. However, it is first appropriate to demonstrate how the friction factor results presented in the earlier part of this chapter can be advantageously used in conjunction with the Gnielinski correlation [55]. The insufficiency of the Gnielinski model in the low-Reynolds-number end of the transition regime, Figure 21, may be conjectured to be due to the friction factor input information that that model requires. To illuminate this hypothesis, it is relevant to examine the Gnielinski correlation, which is

$$Nu_{fd} = \frac{\frac{f}{8}(\text{Re}-1000)\text{Pr}}{1.00+12.7 \cdot \sqrt{\frac{f}{8}}\left(\text{Pr}^{2/3}-1\right)} \qquad (16)$$

The friction factor correlations available to Gnielinski were restricted to Re > 4000. In this work, friction factors bridging the gap between pure laminar and pure turbulent flows have been presented in Eqs. (10), (11), and (13). It is now proposed to use these friction factors to extend the applicability of the Gnielinski correlation to the lower-Reynolds-number end of the transition regime.

The results obtained from this evaluation are presented in Figures 22-24, which display information for the round pipe of constant cross section, parallel-plate channel, and round pipe with an upstream conical enlargement, respectively. The Prandtl number for these results is 0.7 (air).

Figure 22 is for the round pipe of constant cross section. The figure shows the original Gnielinski recommendation of Eq. (16) along with the new results based on the friction factors from Eq. (10). There are two reference lines in the figure for laminar flow, respectively for the uniform wall temperature (UWT) and uniform heat flux (UHF) boundary conditions. It can be seen that the extended Gnielinski results deviate significantly from the original Gnielinski equation for Reynolds numbers below 4000. Furthermore, the extended-correlation results are in perfect agreement with the laminar flow values whereas the original Gnielinski equation gives a value at Re = 2300 that is greater by a factor of about two compared to the laminar limit. On that basis alone, the new results provide a more reasonable outcome.

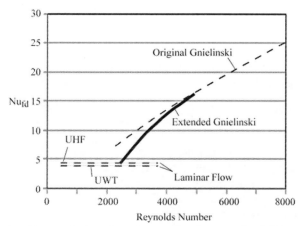

Reproduced with permission from International *Journal of Heat and Mass Transfer,* (in press).

Figure 22. Extended-Gnielinski model Nusselt number predictions for the low-Reynolds-number end of the transition regime. Constant-area round pipe, Pr = 0.7.

Figure 23 is the parallel-plate channel counterpart of the round pipe results conveyed in Figure 22. The information displayed in Figure 23 is qualitatively similar to that of Figure 22, but there are interesting differences in detail. First, the Reynolds number at which the new parallel-plate results coincide with the laminar asymptotes is higher than that for the round pipe. This finding has many precedents. In a comprehensive study of experimental results for rectangular ducts of aspect ratios up to 40, Jones [58] found that the Reynolds number at the breakdown of laminar flow varied substantially with the nature of the inlet, reaching values as high as 7000 before turbulence was established. A similar finding was obtained by the present authors in [31]. In addition, the Reynolds number range covered by the new results is wider than that for the round pipe.

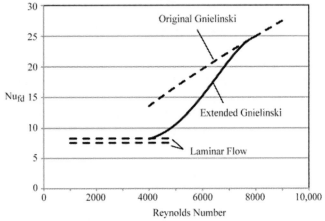

Reproduced with permission from *International Journal of Heat and Mass Transfer,* (in press).

Figure 23. Extended-Gnielinski model Nusselt number predictions for the low-Reynolds-number end of the transition regime. Parallel-plate channel, $Pr = 0.7$.

The last case to be considered is the downstream end of a round pipe fitted with a conical enlargement (diverging nozzle) at its upstream end. The results for this case are presented in Figure 24 in the same format as for the prior figures. A comparison of the results of Figs. 22 and 24 indicates a close but not exact correspondence, although the two situations are both fully developed flows in round pipes. The slight differences are due to the upstream histories of the respective fully developed flows. In particular, the enlargement of the cross section upstream of the fully developed region is remembered by the flow.

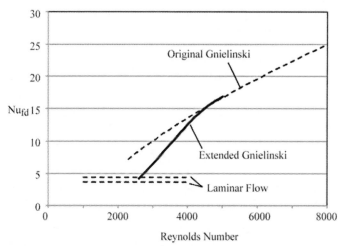

Reproduced with permission from *International Journal of Heat and Mass Transfer*, (in press).

Figure 24. Extended-Gnielinski Nusselt number predictions for the low-Reynolds-number end of the transition regime. Round pipe with an upstream conical enlargement, Pr = 0.7.

The results of the numerical simulations from Eqs. (1) – (9) for the round pipe of constant cross-sectional area have been superimposed on Figure 22. The superimposed results are displayed in Figure 25 as data points. In the figure, squares denote the uniform wall temperature boundary condition, and triangles represent uniform heat flux. Careful inspection of the figure reveals excellent agreement between the Nusselt number correlation of Figure 22 and the results of the simulations. Not unexpectedly, the Nusselt numbers for the uniform heat flux case are greater than those for uniform wall temperature. The Nusselt number correlation threads itself between the results for the two boundary conditions. The level of agreement displayed in Figure 25 provides strong support for the new low-Reynolds-number, transition-regime correlation provided in Figure 22.

A comparison similar to that of Figure 25 is presented in Figure 26 for the case of the downstream flow in a round pipe which is fitted with a conical enlargement at its upstream end. As before, the uniform heat flux results are higher than those for uniform wall temperature, but the gap between the two sets of data is smaller. These simulation results are compared with the new Nusselt number correlation displayed in Figure 24. Once again, there is excellent agreement between the new correlation and those of the simulation, thereby adding further support to the method underlying the obtainment of the new results. The data presented in Figure 25 were extracted from [59].

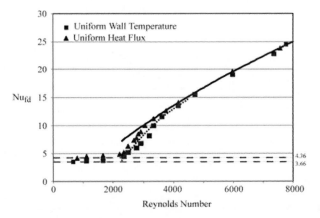

Figure 25. Comparison of the new Nusselt number correlation for the low-Reynolds-number transition regime with those of the numerical simulations for the round pipe of constant cross section and for $Pr = 0.7$.

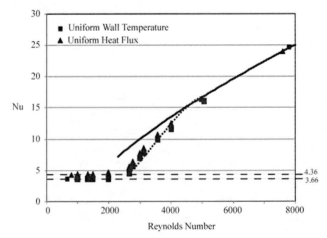

Figure 26. Comparison of the new Nusselt number correlation for the low-Reynolds-number transition regime with those of the numerical simulations for downstream flow in a round pipe with a conical diffuser at its upstream end and for $Pr = 0.7$.

*Chapter 7*

# CONCLUDING REMARKS

Engineering practice is replete with applications where fluid flows in the transition regimes between laminar and turbulent flow prevail. Such inter-regime processes are encountered not only when a laminar flow is transisting into a turbulent flow but also when a turbulent flow is transisting into a laminar flow. The former transition may be termed turbulentization while the latter may be designated as laminarization. Although not widely acknowledged, flows may exist permanently in a transition regime. An example of such a situation is a pipe flow having Reynolds numbers between 2300 and 4500. Heat exchangers are notorious for their frequent operation in a transition regime.

From the standpoint of energy-efficient engineering design, the relevant quantities are friction factors and heat transfer coefficients. Definitive information on the numerical values of these quantities in regimes of flow transition is noteworthy by its absence, both from the standpoints of numerical simulation and experimentation. The work reported here has provided numerical-simulation-based predictions of these quantities. In addition, an available empirically based correlation of heat transfer results has been extended into the low-Reynolds-number end of the laminar-to-turbulent transition regime.

A major issue, treated tentatively here but needing further development, is a logic-based relationship between the turbulent viscosity and the turbulent thermal conductivity.

# REFERENCES

[1] S.S. Varghese, S.H. Frankel, and P.F. Fischer, Modeling Transition to Turbulence in Eccentric Stenotic Flows, *J. Biomechanical Engineering*, 130, paper 014503, (2008).

[2] A. Scotti and U. Piomelli, Numerical Simulation of Pulsating Turbulent Channel Flow, *Physics of Fluids*, 13, 1367-1384, (2001).

[3] G. Alfonsi, Reynolds-Averaged Navier-Stokes Equations for Turbulence Modeling, *Applied Mechanics Reviews*, 62, paper 040802, (2009).

[4] F.P.P. Tan, S. Soloperto, S. Bashford, N.B. Wood, S. Thom, A. Hughes, and X.Y. Xu, Analysis of Flow Disturbance in a Stenosed Carotid Artery Bifurcation Using Two-Equation Transitional and Turbulence Models, *J. Biomechanical Engineering*, 130, paper 061008, (2008).

[5] M.M. Rai and P. Moin, Direct Numerical Simulation of Transitional and Turbulence in a Spatially Evolving Boundary Layer, *J. of Computational Physics*, 109, 169-192, (1993).

[6] M. Alam and N.D. Sandham, Direct Numerical Simulation of Short Laminar Separation Bubbles with Turbulent Reattachment, *J. Fluid Mechanics*, 403, 223-250, (2000).

[7] D.C. Chu and E.M. Karniadakis, A Direct Numerical Simulation of Laminar and Turbulent Flow over Riblet-Mounted Surfaces, *J. Fluid Mechanics*, 250, 1-42, (1993).

[8] U. Rist and H. Fasle, Direct Numerical Simulation of Controlled Transition in a Flat-Plate Boundary Layer, *J. Fluid Mechanics*, 298, 211-248, (1995).

[9] R.D. Joslin, C.L.Strett, and C.L. Chang, Spatial Direct Numerical Simulation of Boundary-Layer Transition Mechanisms: Validation of

PSE Theory, *Theoretical and Computational Fluid Dynamics*, 4, 271-288, (1993).

[10]  O. Reynolds, An Experimental Investigation of the Circumstances Which Determine Whether the Motion of Water Shall be Direct or Sinuous, and of the Law of Resistance in Parallel Channels, *Phil. Trans. Roy. Soc. Lond*, 174, 935-982, (1883).

[11]  H. Emmons, The Laminar-Turbulent Transition in a Boundary Layer - Part 1, *J. Aerospace Science*, 18, 490-498, (1951).

[12]  M. Mitchner, Propagation of Turbulence from an Instantaneous Point Disturbance, *J. Aeronautical Sciences*, 21, 350-351, (1954).

[13]  D. Dhawan and R. Narasimha, Some Properties of Boundary Layer Flow During Transition from Laminar to Turbulent Motion, *J. Fluid Mechanics*, 3, 418-436, (1958).

[14]  P. Libby, On the Prediction of Intermittent Turbulent Flows, *J. Fluid Mechanics*, 62, 273-295, (1975).

[15]  P. Libby, Prediction of the Intermittent Turbulent Wake of a Heated Cylinder, *Physics of Fluids*, 19, 494-501 (1976).

[16]  V. Patel and G. Scheuerer, Calculation of Two-Dimensional Near and Far Wakes, *AIAA J.*, 20, 900-907, (1982).

[17]  R. Mayle, The Role of Laminar-Turbulent Transition in Gas Turbine Engines, *J. Turbomachinery*, 113, 509-537, (1991).

[18]  Y. Suzen and P. Huang, Modeling of Flow Transition Using an Intermittency Transport Equation, *J. Fluids Engineering*, 122, 273-284, (2000).

[19]  Y. Suzen, G. Xiong, and P. Huang, Predictions of Transitional Flows in Low-Pressure Turbines Using Intermittency Transport Equation, *AIAA J.*, 42, 254-266, (2002).

[20]  Y. Suzen and P. Huang, Predictions of Separated and Transitional Boundary Layers Under Low-Pressure Turbine Airfoil Conditions Using an Intermittency Transport Equation, *J. Turbomachinery*, 125, 455-464, (2003).

[21]  Y. Suzen and P. Huang, Comprehensive Validation of an Intermittency Transport Model for Transitional Low-Pressure Turbine Flows, *42$^{nd}$ Aerospace Sciences Meeting and Exhibit*, Reno, NV, January 5-8, (2004).

[22]  F. Menter, T. Esch, and S. Kubacki, Transition Modelling Based on Local Variables, *5$^{th}$ Int. Symposium on Engineering Turbulence Modeling and Measurements*, Mallorca, Spain, (2002).

[23]  F. Menter, R. Langtry, S. Likki, Y. Suzen, P. Huang, and S. Volker, A Correlation–Based Transition Model Using Local Variables, Part I –

Model Formulation, *Proceedings of ASME Turbo Expo Power for Land, Sea, and Air*, Vienna, Austria, June 14-17, (2004).

[24]  F. Menter, R. Langtry, S. Likki, Y. Suzen, P. Huang, and S. Volker, A Correlation –Based Transition Model Using Local Variables, Part II – Test Cases and Industrial Applications, *Proceedings of ASME Turbo Expo Power for Land, Sea, and Air*, Vienna, Austria, June 14-17, (2004).

[25]  J.P. Abraham, J.C.K. Tong, and E.M. Sparrow, Breakdown of Laminar Pipe Flow into Transitional Intermittency and Subsequent Attainment of Fully Developed Intermittent or Turbulent Flow, *Numerical Heat Transfer Part B*, 54, 103-115, (2008).

[26]  B.E. Launder and D.B. Spalding, Numerical Computation of Turbulent Flows, Academic Press, London, (1972).

[27]  B.E. Launder and D.B. Spalding, Mathematical Modeling of Turbulence, *Computer Methods in Applied Mechanics and Engineering*, 3, 269-289, (1974).

[28]  K. Hanjalic and S. Jakirlic, Closure Strategies for Turbulent and Transitional Flows, eds. B. Launder and N. Sandham, Cambridge University Press, (2002).

[29]  J. Smagorinsky, General Circulation Experiments with Primitive Equations, I. The Basic Experiment, *Month Weather Review*, 91, 99-164, (1963).

[30]  J.P. Hartnett, J.C.Y. Koh, and S.T. McComas, A Comparison of Predicted and Measured Friction Factors for Turbulent Flow through Rectangular Ducts, *J. Heat Transfer,* 84, 82-88, (1962).

[31]  W.J. Minkowycz, J.P. Abraham, and E.M. Sparrow, Numerical Simulation of Laminar Breakdown and Subsequent Intermittent and Turbulent Flow in Parallel Plate Channels: Effects of Inlet Velocity Profile and Turbulence Intensity, *Int. J. Heat Mass Transfer*, 52, 4040-4046, (2009).

[32]  G.S. Beavers, E.M.Sparrow, and J. Lloyd, Low-Reynolds-Number Turbulent Flow in Large Aspect Ratio Rectangular Ducts, *J. Basic Engineering*, 93, 296-299, 1971.

[33]  C.F. Colebrook, Turbulent Flow in Pipes with Particular Reference to the Transition Between Smooth and Rough Pipe Laws, *J. Inst. Civil Engineering London*, 11, 133-156, (1938).

[34]  J.P. Abraham, E.M. Sparrow, J.C.K. Tong, and D.W. Bettenhausen, Internal Flows Which Transist from Turbulent Through Intermittent to Laminar, *Int. J. Thermal Sciences*, 49, 256-263, (2010).

[35]  E.M. Sparrow, J.P. Abraham, and W.J. Minkowycz, Flow Separation in a Diverging Conical Duct: Effect of Reynolds Number and Divergence Angle, *Int. J. Heat Mass Transfer*, 52, 3079-3083, (2009).

[36]  J.R. Seume and T.W. Simon, Oscillating Flow in Stirling Engine Heat Exchangers, *Proc. Intersociety Energy Conversion Engineering Conference*, 533-538, (1986).

[37]  S.S. Varghese and S.H. Frankel, Numerical Modeling of Pulsating Turbulent Flow in Stenoic Vessels, *J. Biomech. Eng.*, 125, 445-460, (2003).

[38]  D.C. Winter and R.M. Nerem, Turbulence in Pulsatile Flows, *Annals Biomedical Eng.*, 12, 357-369, (1984).

[39]  E.L. Yellin, Laminar-Turbulent Transition Process in Pulsatile Flow, *Circulation Research*, 19, 791-804, (1966).

[40]  S.S. Varghese, S.H. Frankel, and P.F. Fischer, Direct Numerical Simulation of Stenotic Flows. Part 1. Steady Flow, *J. Fluid Mech.*, 582, 253-280, (2007).

[41]  S. Einav and M. Aokolov, An Experimental Study of Pulsatile Pipe Flow in the Transition Range, *J. Biomech. Eng.*, 115, 404-411, (1993).

[42]  R. D. Lovik, J. P. Abraham, W. J. Minkoqycz, and E. M. Sparrow, Laminarization and Turbulentization in a Pulsatile Pipe Flow, *Numerical Heat Transfer Part A*, 56, 861-879, (2009).

[43]  W. Kays, Turbulent Prandtl Number - Where Are We? *J. Heat Transfer*, 116, 284-295, (1994).

[44]  J. Crimaldi, J. Koseff, and S. Monismith, A Mixing-Length Formulation for the Turbulent Prandtl in Wall-Bounded Flows with Bed Roughness and Elevated Scalar Sources, *Physics of Fluids*, 18 paper number 095102, (2006).

[45]  K. Bensayah, A. Benchatti, M. Aouissi, and A. Bounif, Scalar Turbulence Model Investigation with Variable Turbulent Prandtl Number Applies in Hot Axisymmetric Turbulent Round Jet, *Heat and Tech.*, 25, 49-56, (2007).

[46]  L. Chua, Y. Li, and T. Zhou, Measurements of a Heated Square Jet, *AIAA J.*, 42, 578-588, (2004).

[47]  A. Grachev, E. Andreas, C. Fairall, P. Guest, and O. Persson, On the Turbulent Prandtl Number in the Stable Atmospheric Boundary Layer, *Boundary-Layer Meteorology*, 125, 329-341, (2007).

[48]  P. Bass, S. de Roode, and G. Lenderink, The Scaling Behaviour of a Turbulent Kinetic Energy Closure Model for Stably Stratified Conditions, *Boundary-Layer Meteorology*, 127, 17-36, (2008).

[49] S. Jenkins, J. von Wolfersdorf, B. Weigand, T. Roediger, H. Knauss, and E. Kraemer, Time-Resolved Heat Transfer Measurements on the Tip Wall of a Ribbed Channel Using a Novel Heat Flux Sensor - Part II: Heat Transfer Results, *ASME Turbo Expo 2006, Power for Land, Sea, and Air*, Barcelona, Spain, May 6-11, (2006).

[50] S. Lau, PhD Thesis, Effect of Plenum Length and Diameter on Turbulent Heat Transfer in a Downstream Tube and on Plenum-Related Pressure Loss, University of Minnesota, (1980).

[51] L. Bosmans, MS Thesis, Effect of Nonaligned Plenum Inlet and Outlet on Heat Transfer in a Downstream Tube and on Pressure Drop, University of Minnesota, (1981).

[52] R. Kemink, MS Thesis, Heat Transfer in a Tube Downstream of a Fluid Withdrawal Branch, University of Minnesota, (1977).

[53] D. Wesley, PhD Thesis, Heat Transfer in a Pipe Downstream of a Tee, University of Minnesota, (1976).

[54] A. Black III, PhD Thesis, The Effect of Circumferentially-Varying Boundary Conditions on Turbulent Heat Transfer in a Tube, University of Minnesota, (1966).

[55] V. Gnielinski, New Equations for Heat and Mass Transfer in Turbulent Pipe and Channel Flow, *Int. Chem. Eng.*, 16, 359-367, (1976).

[56] S. Churchill, Comprehensive Correlating Equations for Heat, Mass, and Momentum Transfer in Fully Developed Flow in Smooth Tubes, *Ind. Eng. Chem. Fundam.*, 16, 109-116, (1977).

[57] S. Churchill, A Reinterpretation of the Turbulent Prandtl Number, *Ind. Eng. Chem. Res.*, 41, 6393-6401, (2002).

[58] O.C. Jones, An Improvement in the Calculation of Turbulent Friction in Rectangular Ducts, *J. Fluids Engineering*, 98, 173-181, (1976).

[59] J.P. Abraham, E.M. Sparrow, and J.C.K. Tong, Heat Transfer in All Pipe Flow Regimes - Laminar, Transitional/Intermittent, and Turbulent, *Int. J. Heat Mass Transfer*, 52, 557-563, (2009).

# INDEX